This book belongs to

_____

_____

Dear Fellow Puzzle Enthusiast,

Thank you for your purchase of the "Large Print Sudoku Puzzle Book with Solutions - Easy to Hard". We're delighted to have you embark on this journey with one of our carefully crafted collections.

We are a small family-run business driven by passion, and know that each puzzle is crafted to offer not just a challenge but a stimulating experience.

As a small business, your support is invaluable to us. Consider sharing your experience by leaving us a review—it's more than just feedback; it's a vote of confidence that makes a significant impact on our small business journey.

Thank you for being part of our community. Happy puzzling!

SUDOKU

PUZZLES FOR ADULTS

				6			2	
9		7	3	8			4	
			4			5		3
	6	9		5		4		
					1			7
2		5				6		
5								
	2							
	4	8		3		9		

120 easy to hard puzzles

LARGE PRINT

How to play Sudoku

The rules of Sudoku are simple and easy. It is precisely that simplicity that makes finding solutions and solving these puzzles a real challenge.

To play Sudoku, players only need to know the numbers 1 to 9 and be able to think logically. The goal of this game is clear: the task is to complete the grid by filling in the numbers 1 to 9. The difficulty lies in the limits placed on players to fill the grid.

Rule 1: Each row must contain the numbers from 1 to 9, without repetitions

The player must focus on filling each row of the grid, making sure there are no duplicate numbers. The order of the numbers does not matter.

Every puzzle, regardless of the difficulty level, begins with the numbers allocated on the grid. The player must use these numbers as clues to find out which digits are missing in each row.

Rule 2: Each column must contain the numbers from 1 to 9, without repetitions

The Sudoku rules for columns on the grid are exactly the same as for the rows. The players must also fill these with the numbers from 1 to 9, making sure each number occurs only once per column.

The numbers allocated at the beginning of the puzzle work as clues to find out which digits are missing in each column and their position.

Rule 3: The digits can only occur once per block (nonet)

A regular 9x9 grid is divided into 9 smaller blocks of 3x3, also known as nonets. The numbers from 1 to 9 can only occur once per nonet.

In practice, this means that the process of filing rows and columns without duplicate numbers within each block imposes further restrictions on the placement of numbers.

Rule 4: The sum of every single row, column, and nonet must be equal to 45

To find out which numbers are missing from each row, column or block or if there are any duplicates, the player can simply count or sum the numbers.

When the digits occur only once, the total of each row, column and nonet must be 45.

1+2+3+4+5+6+7+8+9=45

Easy Sudoku Puzzles

Puzzle #1

		9				8		
		2			1			
		8	4		9	5	2	6
					6		5	
4	6				3	7		2
2			1					3
8		4	9			2		
5		3	2		8		1	

Puzzle #2

			3		5			6
3				7		8		
				8			2	
	4		1	3	2		7	8
		2		6	7	4		1
7	6		9				3	5
6	2	7				1		
9						5	6	

Puzzle #3

4			2	9	1		6	
2		1		3		5	4	
			4	5	6		7	
8	2				3	9		7
			1	2				
	1					6		
								1
	7	6	3			2	8	4
	3	2		4			5	

Puzzle #4

				4		2	8	5
1				2			3	
	5		3					
		1		3	7	4		
3			8	9		7		1
				1		9		
		3		7		8		2
	4			5			9	6
		8						

Puzzle #5

				4	7		9	5
9	8							
			8		2		4	
1				8		2	5	
	7	9		5	3	1		
					1	7		
4	9	8	3	7	6			
		1			4			
7		2						6

Puzzle #6

							8	
7	9	1			6		5	4
			1					
	8		4				3	6
	5	4				8	7	
	1	2		3	8	5	4	
		3		1		9	2	
1	7		3			4		8
		8	6	9				

Puzzle #7

	3				8		7	2
6			3	2				4
	9	7	1		6			
9			8	3				7
8	4					9		1
				1	9		4	
4			7	6		3	1	
	2			9				
	1		4			7		9

Puzzle #8

	4			5	2			
6	3	2		4		7		8
8	9			7	6		4	
3			5		4	1		9
		4						5
	6	9			3			2
7			2		8	9		
						3		
					7	5	8	

Puzzle #9

3		9		4	2			
	4	2						
1	6			3			7	
9						6		
			6	1	3		5	8
		5	2	9		1		
				2	6		1	5
	3		9					
	2	1	8					

Puzzle #10

6		3				1		5
		5		1			7	
1				8		3		
7						8	2	3
	3	6		2			1	4
5	2		8		3		6	
	6					7		
						4		2
4	1	7			5			

Puzzle #11

	5	6	1		8		2	
	7		9					
			7			1	5	
		3	4		7	5	9	
1							8	6
	4		2				3	
		5		2	4			9
6					9		4	8
		8		3		2	7	

Puzzle #12

		1		7		3	6	8
	2	8					4	9
6				4	8		2	
			9					
				5				
8	3					2	5	
	1	6	7					
		4	1	3				5
			5	8	6	9		4

Puzzle #13

9					4		3	5
	4	1	9		3			
7			6	8		1		9
8			3	6				
4	7			2			8	
			4		8	9	6	1
1	8		5					7
			2		1	4		
	9	2			7			6

Puzzle #14

					2		8	
4			9					5
	5	1	8				3	
9		8	3	7				
		7	5		4	1		
5			1	2		3		
8				9		5	6	3
			6			2		
	9	3				8		

Puzzle #15

				3		6		1
8		4	7	1			9	
	9				4		3	
3					2		5	
2			3	6				9
	4			9			2	
	6		2					
9	2		6			1		
5		8	1	4		2		

Puzzle #16

			6		9		3	
	6			3	4		7	8
	8		2			1		
						9		
			7	4	8	2	5	6
				1	6			
		7	5	9	1		2	
	9			6	2	3		
		1	4	7		8		5

Puzzle #17

		8						
6				9			1	
	1	4	2					9
		5	4		2		8	6
					8	2	4	
					1	5		
2	7	9		4	3	6	5	8
5		3	8			1		
	4	1	7				9	2

Puzzle #18

		4		7	9	3	2	
8	2				4	1	7	
			3		8			
	1	6		9	3	7		
9		8	7					
7				6		4		
	7	1	9				5	
3		5	2				1	6
	6			3				

Puzzle #19

9	4			5		6		
1	8	5		4				7
2							8	
					3	2		
	5		6			4		
		6			2		5	9
6			8	1				5
4		1						
	7			6			9	2

Puzzle #20

			2		5		7	4
	7		4	3	8		2	
					6			
	3		8					9
	1	4		7			5	2
		8		2	4	7		3
	6	9			2	1	3	
7			3			6	4	
				6	7			

Puzzle #21

		3				5		
	1		3				9	8
	6			5				2
	3		9				2	
	4	9			5	3		
7	8		4	3			6	5
							8	
3			2			6		9
6	9		7					

Puzzle #22

7	1	3		6		9		
5				3	4			
			7				5	
		1		4	5			3
8	2					7		
	3	4				6		5
2	9	6			3		7	
1					9			8
					1			

Puzzle #23

1		8		7		5		
		6	8		5			4
		4				7	9	
		7						2
9	8			1		4		
2		1		4			5	3
4			1					
	7	9	5		4	6	2	
		3						

Puzzle #24

		3	9					4
		5				2	7	
		7	4					
		1	9	2		6	5	
	4	9						
			3		7			
3		4				5	8	
	1			3			2	
	5		6		8	4		

Puzzle #25

				5	4			
		1	8			5	2	
6			7					
	3			6				7
5		2					3	6
	6	5		2	7	3		9
	2	8	4				7	
4			5		3	8		

Puzzle #26

	8		4		6	5		
2			7		1			
		7		3				1
		2	9					
9		8						4
			8		3	1		2
4								
8			2	9				
			6			8	4	7

Puzzle #27

						4		1
8						7	5	6
			5		7		9	
		4	6	9		3	1	
6			8			9		
3				1	4			
								8
	3			4		5	7	
		8	3	5	9			2

Puzzle #28

4	7						6	1
				6				
			5		2	4		
						3		5
7			6			2	4	
					4			
9	5	7	3		8		2	
1	3						7	
6		4		9				

Puzzle #29

	4					3		7
	9							
	6			9	3	4		
2	8		1		9			
			2				4	
5		9	3	7			2	
			4	3		8		
4				1			5	9
		8		2	7	1		4

Puzzle #30

4		8					6	7
			9		1	3	8	
1			8				9	4
7				3				2
9	6					5		
				2		9	7	
					2		4	1
	1						2	
2					7	8	5	

Puzzle #31

5		3	8					
	7	2	3			1		6
		4						7
3	2	6		4		9		5
	5			3				
	9		6	2			4	
2		7		6				9
				7		5	6	8
	4	5	9	8			2	1

Puzzle #32

		4	3	6				
9	2				4			6
5		1						
	1			2	5			
4				1				7
				3	7	2		4
						9		3
	5		7			4	6	
1		9					8	2

Puzzle #33

3		4		5	2			6
				3		7		8
							3	
9	2	7				6		3
		1		9			4	
		6			1			9
	9	8						
6		2				4		1
7			6		8			

Puzzle #34

		8				7	2	5
		5	9					3
		3		7			1	
4	8			3			9	6
					2			4
	3	2		9		5		
5		9	8	6	1			
			4					
			5		9			

Puzzle #35

			2			7	3	4
			9	7	3			
	2				1		8	
5	3				4			2
								9
	8	4						
		2		6	9			
6	1	7	3			4		
3				1	7	2	6	

Puzzle #36

4		1	9				3	
	3				5			9
7			2		3	1		4
			3		6	8		
9				7				
	6	8					1	
1		7						
2						6	5	
6		5		3	8	2		1

Puzzle #37

		4	5		1		7	
	8			4		5		9
			8			2	5	
1	2		3					
7				5	2	1	9	
5		6			3	8		4
	1	3			4			
						9		1

Puzzle #38

7	2							
			5	7				
	3		9	6			2	
	4	2						1
3	7		2	1	8	5		
				4			3	
					5		1	7
2		7		3		4		6
1	9		6					

Puzzle #39

		5		8			9	
6								
7	2				9	1		4
			4		1			8
1		9			3	4		7
		7		6				
			2					
9	1		7			6		2
5						7	4	

Puzzle #40

6	2	1		5				7
					4	5		
8			7	2		1		9
		6		3				2
7		3	2					
1							7	
	7							
			3	7	2	8	9	6
2		9	4		8		5	

Medium Sudoku Puzzles

Puzzle #41

8						3		
	2		9			5		
			6		5		2	
	7	5		6				1
	6				7			
9		1		5		2	6	
	3	8			2	6		
	1						4	5
			8					

Puzzle #42

					2			3
							4	8
				1		5	7	
		1		5		7	6	4
9				8				2
6				4				
4								
2	9				8			1
5	6	8			7			

Puzzle #43

4	5		8					
				5				
9			6			3		
	2			8	6			7
						4		
	6	4	1	2	3			
1			2	6	8	7		
6	9	8			7		5	
		7	5					6

Puzzle #44

8	7			1	5			
2	4					6		
			3	4			7	
				2		4	8	
9					7		2	
	9							
		5	2				4	
1	8			6	4	2	3	

Puzzle #45

				8		4		
		6	9		2			
7				4				
	8			7				1
						8	3	5
	9			6				
3	7				9		4	
8		2			3	9	1	
6						3		

Puzzle #46

7		4		3			6	9
	1		9			5		7
				4				
	5		7			9		
					8			
6		8	4			3	7	
2	4	7	3	9				
	9							2
					5			

Puzzle #47

	5		4					
9				2		7		8
				3			6	
5		1				3		
		3	5		4		8	
	2				7		4	1
			1			8		6
4	6				2	9		
1								

Puzzle #48

		4			8	1	7	5
	3	1	9			6		
6					5			3
3				7	1		5	
1								4
		5		3				
				9	8			
	8		5				4	
9			3				2	

Puzzle #49

						7		
		8	2					4
			7		9	1		
	6	1		2				7
8		4		6			2	1
7				3	4			
4			5	9			7	
	1	2	4	7	3		5	
								3

Puzzle #50

1							4	
			2	9		5	8	6
		8		3		7	1	
	1			6	8			3
9			5					8
3				2		6		
8		6					5	
5	7	9				2	3	
					2			

Puzzle #51

8				6			2	7
	3				7	9		
							1	
			6					
		1		5	4	2		8
4		5						
			2				6	
3				8	1			4
7							3	

Puzzle #52

	2				3			7
7				8			9	
6		1		9	7		5	
9				3				
	4		1		2			9
	6	2						3
4								5
	9		3		8	2	1	
						6		

Puzzle #53

		5	3		9	2		
		1		2				5
	3		6					9
	1	6	4					8
					2		7	1
4					8			
	4	8		3		1		6
			8				4	
6		3	1		5			

Puzzle #54

		2	8				9	6
	4					1		
		8	3	1		2		
7	3	5	6			8	2	
4	2					9		
	1							3
5	8		2	6		7		9
			9					8
3		4	1	7		5	6	

Puzzle #55

	2		5			3		
		5					1	
		3		6	2	8		
	8		2		3			7
4				1		5		
3			7				9	
6		8			1			
					4			
				8		2	4	

Puzzle #56

				4		9	8	
	6	4		1				
			6					2
	2			3	1		6	
		3	4	5			2	
					6			9
2	3		1		7	6		4
4		9					1	
	7					2		3

Puzzle #57

			3			4		
						2	3	
3				8			7	5
	1				4	7		
				7			9	
9		2				6		1
8					1			
6		3						
		7	4	5	3			6

Puzzle #58

		9			3			4
9		8			3	7	5	
			8					
				2				6
			5				9	
3	2	9				1		5
6				1				
	7	3	4	2				
				9			1	7

Puzzle #59

		6	1				5	
				9				
	8						7	2
1			4	2	6		8	3
		7		8		1		
				3		5		
	1	3		4	7			
	6				2			
	7			6	9			8

Puzzle #60

7		2	5		1	3		
4		5				7		8
	1				9			2
3							9	
				7	8	6		
	2		6					
		7			5	2		
2	4	1		6				
		9						

Puzzle #61

						9	7	
		5	7		9	6		
								1
	4	6						2
	5					4	8	
			3	5	1			
	9			7	8			4
8			1				6	9
	7						2	5

Puzzle #62

			6	7		2		
	7	9		2	4	6		
				8	1			
9	4				3		5	
6		5		1	2			
		2	4					
		4						
		1	5			3	7	4
		8				9		1

Puzzle #63

9	3		4		6			
	7				5	1	8	
2	5	1		9				
	6						3	8
		9	8					
	4			7	3	9		
	9	6		8	4	2		5
		3		2			7	
5					7	3		4

Puzzle #64

7			6				4	
			2			5	7	
6	3	1						
9					6			
		4	3	7	5			8
	1							
8	4	9				1	5	
	7		8	4			9	
								7

Puzzle #65

			2			5		
8			4				7	
5	4		1			2		6
				7			4	
					4		1	7
7						9	5	
4		7			6			
	2	9	7					
6			5				9	

Puzzle #66

	8		2	4		7		
	7			5	6		4	
						9		
		6			2		8	
		1	5	3				4
						5		
		9	3		8			
	5				7			6
	1	3	6				9	

Puzzle #67

	8		9	4		2		6
7				6				
2	6		3	1				
6		9		3			7	
	3						1	4
		5			1			9
				7			5	
						8		
			5			4		

Puzzle #68

		6			5		9	3
2				6				
							2	
				7	2			
3	9	2			1			7
1								
6	5		7	4			3	
		1	5	2	3	4		
				1		8		

Puzzle #69

	1	9	5	6	2			7
		4			8		1	
						5		
1	3		6			9		
						4	8	6
		5						
4			2		9			
		3			1		2	9
						3		

Puzzle #70

		2			9			6
				3	2	9		
	1			5	6		4	
7			3			1	9	
		6			5			2
5			2					
3		8		2			5	
	5			4	3			7
				8				

Puzzle #71

		5	2					
3					4	9		6
		9					1	
8				6		2		1
5							8	
	4			8		6		
				2				
	1			4			7	3
	7				1	8		

Puzzle #72

	7						2	3
				9		7		8
			5					
	1			6	9	8	3	
5								4
	6	4		5				
		9				4		
1			8	3			5	2
2		8		7				

Puzzle #73

	9	3			5			
			7		2			
	5		3	9		1		
	1	5		6		8		
9	7						6	
2								1
				4			3	
	6				3		5	
		9			8			4

Puzzle #74

8					1	2		4
	2	4		9			7	1
		5		2		6		
	5				4	3		6
9		6						
		3		2	6	7	8	
			3	8			9	
					7	1		

Puzzle #75

5			6			9		
	3					6	1	
		6					8	
	4		2		6			1
	8							
		5	4		1	3	2	9
7				9				
2						1		3
	6				3		7	2

Puzzle #76

	4					6		1
	1		2	4	9	7		
			1					8
	6	2		7				
	7	8	9		3			
	3		8	5				
	2	3		8			6	
			3					7
1								

Puzzle #77

3	8				9	2		1
2		6		4	1	3		8
4		5						
								6
				8			7	
		8	5				3	
		1		7		5	2	
5						1		4
	3			2	5	7		

Puzzle #78

6				3	1			
		1				2		
	5	7			6			1
	8	6			5		2	
5	4						8	
1					2			5
4				2			7	
			7		9	5		2
2				6				

Puzzle #79

9						1		
2		7				5		
			5	7		3		6
	1		4			9		
6		2		9				
3					2	7		
		6				4		
		9		2	4		5	7
			1			2	3	

Puzzle #80

							6	4
6	4						5	
9	7				1	2		
				9	6			
1		6		8				
		7					3	8
5							7	
			9	2	8			
	6			7		9	4	

Hard Sudoku Puzzles

Puzzle #81

	4		2	9	5	3		
1				7				5
	3				4			9
						8		
2		3		4		5		
9			6				1	
	5					9	6	
	2	7			3			
	6			8				

Puzzle #82

						5		8
	9			5				
	2				3	6		
		5	6		8			
		4					5	
			9				7	3
4	3				9			
		8	3		6	2		9
			7					

Puzzle #83

	2		4			7		
						8		2
4	8	3			5			
	1				9			3
		7		5		4		1
							8	
			6	3				
5	4				1			
	9	8	5			1	2	

Puzzle #84

			6			9		
	1		8		4			
								5
				4	8			6
9	2			1	3		7	
5	8		7				1	
2	3	1						
						2	8	4
		5			7		6	

Puzzle #85

3	5		8				6	4
1	8							9
					1			
5		1				3		
6		8		2			1	
							7	
					6	8	5	
	4		3	7				
			1				3	

Puzzle #86

		8	2	9	4		6	
			1	8	6			
4								
				1		8	3	
		4						
	6	5						
7		2		5				6
			8		7	9	2	
5						1	8	

Puzzle #87

	3		7					9
		7	9				4	5
					2		6	7
	8	2					1	
3	5					6	9	
6			3			5	8	
	6		1	7				
				8	4			
			2				5	

Puzzle #88

	2	1	7		4			
8	5	6		1	3			
1						2		
				6			7	4
			4	3			5	
		7					9	
2	4	5	9					8
		8		6			2	7

Puzzle #89

	3				8			
5		9		2			3	1
				4		8		
						6		3
		7				2	8	9
	1	2	9		3	5		
				9				
6							5	
			5		7		6	

Puzzle #90

				4	6			
	2	7			9	1		
				2	1		3	
1	6							3
	5					2	4	
4			5			9		
					2		7	
		9		3	5			1
	3							

Puzzle #91

		9			8			5
	2	8	3				7	
3			4		5			
			2					3
9					4	8		
	3	1				7	6	
				5	1	4		
			8					
6						3	1	

Puzzle #92

2	5		1		7			
		1	9	4				
7			6	5				1
	3		2				9	7
					1	3		5
						2		
5				3		1		4
6		3	5			9		

Puzzle #93

5		8	7				3	1
	2	7	3					
				2				7
	6							
	5					6	7	
	8	2			3	4	1	5
							6	9
	3		9		4			
			1	5				

Puzzle #94

								8
9	4	6		8	1	5	3	
				3	5	7	4	
		9		4			7	
4	7					1		5
	5					3		
	1				9	4		
8				2				
7	9							

Puzzle #95

			4	3		9		
		7			6		1	3
3	2							
	4	2			8	1		
	6		7	2			5	
							2	
1	8		5		7			
						5	6	
			1					4

Puzzle #96

		8	7	6			4	
9	1					8		
				8	1	3	6	
					2		5	
					4	9	2	1
3			5	9	7			
					8			
	8			5		4	3	
7	9			4			1	

Puzzle #97

1						7		9
								2
					7	5	6	8
7						3	2	
		2	1		8			
	9			7	6			
	8			5			4	
	3				4	2		
			9				8	

Puzzle #98

7					3		5	2
			4			3		
		3	6	7				1
	3	8			2			9
				4				5
				9	7			
9			2	3	6		1	
4	1	6				2		
			9			7		6

Puzzle #99

		8			9			
			4				9	
9	1	6		3	8			
			6	2		3		
6		3		4		7	1	
4				1		6		9
5	3	1					8	
							3	
				2				7

Puzzle #100

	4	5				7		
				9		1	6	
		7		8	5			
				4		9		1
				1		5		
	7		8		9	6	2	
	5	8				3		
	9	2	1					
1							9	2

Puzzle #101

9		8	4				2	5
				5			8	
						1		
8		5	7				3	
		6	2	1		9		8
6					2			7
		1	6				5	
4				9	8			6

Puzzle #102

			2	4				1
4			8					3
6		1					8	
8						9	5	
		6						
		4					3	2
					6			9
	7		4	3			2	
2			1		9	5		

Puzzle #103

		8			4	7		
	5			7			8	
4			2				6	
5	3		6	2				
				4	1			
					5			3
		6	5		2	4		
		7						
1		5	7		6			8

Puzzle #104

3		9	5					1
		4		1			6	
							7	
5	6		2		3	1	8	9
	1			5	6			
9	4						5	2
	2	6		7	8			3
	3		8		9	1		

Puzzle #105

		3	6		2			9
2								8
	1			5				
	2	1		6		5		
8						6		
5		4		1			3	
3			1	9				5
						9	8	4
					7			

Puzzle #106

		6	2			4		3
1			8	4		6		
	3			9			5	6
9								
6	5			1	4		7	
	6	3				2		8
					2		4	
4					9	3		

Puzzle #107

9	7	4						
6		8		4			9	
	2	3			1			4
		6	4	3			2	7
		5						1
	4				2	9		
					9	7		
			1					6
7			2	6			8	

Puzzle #108

		9						
	1		8				5	
			9			4	1	3
					9		2	
2	5				7			6
				4		5	3	1
		8			4	6		
	3		7					
		7	5					2

Puzzle #109

				8	1		2	
			2			4		6
		9		5	6		3	
	5						8	
1			9	6				
		2	8					9
		1						
			6	3		7		8
2				1	9	5		

Puzzle #110

		6					2	1
4				2		9		
				3		6		
	6		9			1		3
3		9		1		2	4	
	4		3		6			
		2			1			
			7	9		8		
1		5		8				9

Puzzle #111

8	4	7						
			6	4	8			
	6	9			7		8	
5					3	2	9	
7	2		4	1	9			5
			2		6	4		3
				7			6	
		6		9			5	7

Puzzle #112

4				6		1		
		8			5			
6	3		1		8	4	9	
			8					
	1	7		2				
2					7	3	8	1
				1		9		
			6			2		7
					2			

Puzzle #113

5	1				7	9		6
9								
		3	4				2	
	8			6	3			
	6		1	3				9
					8			
4	3			1				
		6						
	2	5	7					3

Puzzle #114

	4	2			5	3		8
	5	7						6
						2	5	
				4		6	9	1
		1	3			4		
		9				7		
		3	1					7
	7		6					
		5	2		3			

Puzzle #115

8		6						
	2							8
9		4		8			5	
			5			6		1
1	5					9	2	
			4					7
2							9	
		1		5				
5				9	6	2	4	

Puzzle #116

	7	5				9		2
						1	7	6
			2					5
	8							
	9		1	3		4		
					4	3		
6				8				
			4	9			8	
4	1	8		5		6		

Puzzle #117

				5	4			3
8	3	2			1			
			8			4		7
	7	9						6
		6		1	2			9
	2		4					
1		8	5	3		9		
7		3				6		

Puzzle #118

				4			5	1
					9	3		
7						4	2	
1	6		3			2		5
				8	2			
		5						
	5							
		2		1	8		4	9
9					5			3

Puzzle #119

	5			2	6			
3	4			8			5	
	8	9		7		1		
9			2				3	
				1				4
			9		8			1
		2	6					
	1	4		5				2
8	6							

Puzzle #120

	3				1			
		6	8				9	
	9		5		2			
	6					8		3
8				3				
9			7				4	5
	4				9			7
		5			7		2	
2					4	3		

Puzzle Solutions

Puzzle #1

3	4	9	6	5	2	8	7	1
6	5	2	7	8	1	3	4	9
1	7	8	4	3	9	5	2	6
9	3	7	8	2	6	1	5	4
4	6	1	5	9	3	7	8	2
2	8	5	1	7	4	9	6	3
7	2	6	3	1	5	4	9	8
8	1	4	9	6	7	2	3	5
5	9	3	2	4	8	6	1	7

Puzzle #2

2	8	4	3	9	5	7	1	6
3	9	6	2	7	1	8	5	4
1	7	5	6	8	4	3	2	9
5	4	9	1	3	2	6	7	8
8	3	2	5	6	7	4	9	1
7	6	1	9	4	8	2	3	5
4	5	3	7	1	6	9	8	2
6	2	7	8	5	9	1	4	3
9	1	8	4	2	3	5	6	7

Puzzle #3

4	5	7	2	9	1	8	6	3
2	6	1	7	3	8	5	4	9
3	8	9	4	5	6	1	7	2
8	2	4	5	6	3	9	1	7
6	9	5	1	2	7	4	3	8
7	1	3	9	8	4	6	2	5
5	4	8	6	7	2	3	9	1
9	7	6	3	1	5	2	8	4
1	3	2	8	4	9	7	5	6

Puzzle #4

7	3	6	9	4	1	2	8	5
1	8	9	7	2	5	6	3	4
4	5	2	3	8	6	1	7	9
6	9	1	5	3	7	4	2	8
3	2	5	8	9	4	7	6	1
8	7	4	6	1	2	9	5	3
5	6	3	4	7	9	8	1	2
2	4	7	1	5	8	3	9	6
9	1	8	2	6	3	5	4	7

Puzzle #5

2	1	3	6	4	7	8	9	5
9	8	4	1	3	5	6	2	7
5	6	7	8	9	2	3	4	1
1	4	6	7	8	9	2	5	3
8	7	9	2	5	3	1	6	4
3	2	5	4	6	1	7	8	9
4	9	8	3	7	6	5	1	2
6	3	1	5	2	4	9	7	8
7	5	2	9	1	8	4	3	6

Puzzle #6

2	3	6	5	4	9	1	8	7
7	9	1	2	8	6	3	5	4
8	4	5	1	7	3	6	9	2
9	8	7	4	5	1	2	3	6
3	5	4	9	6	2	8	7	1
6	1	2	7	3	8	5	4	9
4	6	3	8	1	7	9	2	5
1	7	9	3	2	5	4	6	8
5	2	8	6	9	4	7	1	3

Puzzle #7

1	3	4	9	5	8	6	7	2
6	5	8	3	2	7	1	9	4
2	9	7	1	4	6	5	8	3
9	6	1	8	3	4	2	5	7
8	4	2	6	7	5	9	3	1
3	7	5	2	1	9	8	4	6
4	8	9	7	6	2	3	1	5
7	2	3	5	9	1	4	6	8
5	1	6	4	8	3	7	2	9

Puzzle #8

1	4	7	8	5	2	6	9	3
6	3	2	9	4	1	7	5	8
8	9	5	3	7	6	2	4	1
3	7	8	5	2	4	1	6	9
2	1	4	7	6	9	8	3	5
5	6	9	1	8	3	4	7	2
7	5	6	2	3	8	9	1	4
4	8	1	6	9	5	3	2	7
9	2	3	4	1	7	5	8	6

Puzzle #9

3	5	9	7	4	2	8	6	1
7	4	2	1	6	8	5	3	9
1	6	8	5	3	9	2	7	4
9	1	3	4	8	5	6	2	7
2	7	4	6	1	3	9	5	8
6	8	5	2	9	7	1	4	3
8	9	7	3	2	6	4	1	5
4	3	6	9	5	1	7	8	2
5	2	1	8	7	4	3	9	6

Puzzle #10

6	4	3	9	7	2	1	8	5
9	8	5	3	1	4	2	7	6
1	7	2	5	8	6	3	4	9
7	9	4	6	5	1	8	2	3
8	3	6	7	2	9	5	1	4
5	2	1	8	4	3	9	6	7
2	6	9	4	3	8	7	5	1
3	5	8	1	6	7	4	9	2
4	1	7	2	9	5	6	3	8

Puzzle #11

3	5	6	1	4	8	9	2	7
2	7	1	9	5	3	8	6	4
9	8	4	7	6	2	1	5	3
8	6	3	4	1	7	5	9	2
1	2	7	3	9	5	4	8	6
5	4	9	2	8	6	7	3	1
7	3	5	8	2	4	6	1	9
6	1	2	5	7	9	3	4	8
4	9	8	6	3	1	2	7	5

Puzzle #12

4	5	1	2	7	9	3	6	8
3	2	8	6	5	1	7	4	9
6	9	7	3	4	8	5	2	1
1	6	5	9	2	3	4	8	7
7	4	2	8	6	5	1	9	3
8	3	9	4	1	7	2	5	6
5	1	6	7	9	4	8	3	2
9	8	4	1	3	2	6	7	5
2	7	3	5	8	6	9	1	4

Puzzle #13

9	2	8	7	1	4	6	3	5
6	4	1	9	5	3	8	7	2
7	5	3	6	8	2	1	4	9
8	1	9	3	6	5	7	2	4
4	7	6	1	2	9	5	8	3
2	3	5	4	7	8	9	6	1
1	8	4	5	3	6	2	9	7
3	6	7	2	9	1	4	5	8
5	9	2	8	4	7	3	1	6

Puzzle #14

7	3	9	4	5	2	6	8	1
4	8	6	9	1	3	7	2	5
2	5	1	8	6	7	9	3	4
9	1	8	3	7	6	4	5	2
3	2	7	5	8	4	1	9	6
5	6	4	1	2	9	3	7	8
8	4	2	7	9	1	5	6	3
1	7	5	6	3	8	2	4	9
6	9	3	2	4	5	8	1	7

Puzzle #15

7	5	2	9	3	8	6	4	1
8	3	4	7	1	6	5	9	2
1	9	6	5	2	4	7	3	8
3	1	9	4	7	2	8	5	6
2	8	7	3	6	5	4	1	9
6	4	5	8	9	1	3	2	7
4	6	1	2	8	3	9	7	5
9	2	3	6	5	7	1	8	4
5	7	8	1	4	9	2	6	3

Puzzle #16

1	7	5	6	8	9	4	3	2
9	6	2	1	3	4	5	7	8
4	8	3	2	5	7	1	6	9
7	4	6	3	2	5	9	8	1
3	1	9	7	4	8	2	5	6
2	5	8	9	1	6	7	4	3
8	3	7	5	9	1	6	2	4
5	9	4	8	6	2	3	1	7
6	2	1	4	7	3	8	9	5

Puzzle #17

9	5	8	6	1	7	4	2	3
6	2	7	3	9	4	8	1	5
3	1	4	2	8	5	7	6	9
1	3	5	4	7	2	9	8	6
7	9	6	5	3	8	2	4	1
4	8	2	9	6	1	5	3	7
2	7	9	1	4	3	6	5	8
5	6	3	8	2	9	1	7	4
8	4	1	7	5	6	3	9	2

Puzzle #18

6	5	4	1	7	9	3	2	8
8	2	3	6	5	4	1	7	9
1	9	7	3	2	8	5	6	4
5	1	6	4	9	3	7	8	2
9	4	8	7	1	2	6	3	5
7	3	2	8	6	5	4	9	1
4	7	1	9	8	6	2	5	3
3	8	5	2	4	7	9	1	6
2	6	9	5	3	1	8	4	7

Puzzle #19

9	4	3	7	5	8	6	2	1
1	8	5	2	4	6	9	3	7
2	6	7	1	3	9	5	8	4
7	9	4	5	8	3	2	1	6
8	5	2	6	9	1	4	7	3
3	1	6	4	7	2	8	5	9
6	2	9	8	1	7	3	4	5
4	3	1	9	2	5	7	6	8
5	7	8	3	6	4	1	9	2

Puzzle #20

9	8	6	2	1	5	3	7	4
1	7	5	4	3	8	9	2	6
3	4	2	7	9	6	5	8	1
2	3	7	8	5	1	4	6	9
6	1	4	9	7	3	8	5	2
5	9	8	6	2	4	7	1	3
8	6	9	5	4	2	1	3	7
7	2	1	3	8	9	6	4	5
4	5	3	1	6	7	2	9	8

Puzzle #21

8	2	3	1	9	7	5	4	6
4	1	5	3	2	6	7	9	8
9	6	7	8	5	4	1	3	2
5	3	6	9	7	1	8	2	4
2	4	9	6	8	5	3	1	7
7	8	1	4	3	2	9	6	5
1	7	2	5	6	9	4	8	3
3	5	4	2	1	8	6	7	9
6	9	8	7	4	3	2	5	1

Puzzle #22

7	1	3	5	6	8	9	4	2
5	6	2	9	3	4	1	8	7
4	8	9	7	1	2	3	5	6
6	7	1	2	4	5	8	9	3
8	2	5	3	9	6	7	1	4
9	3	4	1	8	7	6	2	5
2	9	6	8	5	3	4	7	1
1	4	7	6	2	9	5	3	8
3	5	8	4	7	1	2	6	9

Puzzle #23

1	9	8	4	7	2	5	3	6
7	3	6	8	9	5	2	1	4
5	2	4	3	6	1	7	9	8
3	4	7	6	5	9	1	8	2
9	8	5	2	1	3	4	6	7
2	6	1	7	4	8	9	5	3
4	5	2	1	8	6	3	7	9
8	7	9	5	3	4	6	2	1
6	1	3	9	2	7	8	4	5

Puzzle #24

5	2	3	8	9	7	1	6	4
4	9	8	5	6	1	2	7	3
1	6	7	3	4	2	8	9	5
7	3	1	9	2	4	6	5	8
6	4	9	7	8	5	3	1	2
2	8	5	1	3	6	7	4	9
3	7	4	2	1	9	5	8	6
8	1	6	4	5	3	9	2	7
9	5	2	6	7	8	4	3	1

Puzzle #25

2	8	6	3	5	4	7	9	1
9	5	3	2	7	1	6	8	4
7	4	1	6	8	9	5	2	3
6	1	9	7	3	2	4	5	8
8	3	4	9	6	5	2	1	7
5	7	2	1	4	8	9	3	6
1	6	5	8	2	7	3	4	9
3	2	8	4	9	6	1	7	5
4	9	7	5	1	3	8	6	2

Puzzle #26

1	8	3	4	2	6	5	7	9
2	9	5	7	8	1	4	6	3
6	4	7	5	3	9	2	8	1
5	1	2	9	4	7	6	3	8
9	3	8	1	6	2	7	5	4
7	6	4	8	5	3	1	9	2
4	5	1	3	7	8	9	2	6
8	7	6	2	9	4	3	1	5
3	2	9	6	1	5	8	4	7

Puzzle #27

7	5	2	9	6	3	4	8	1
8	9	3	4	2	1	7	5	6
4	6	1	5	8	7	2	9	3
5	8	4	6	9	2	3	1	7
6	1	7	8	3	5	9	2	4
3	2	9	7	1	4	8	6	5
9	4	5	2	7	6	1	3	8
2	3	6	1	4	8	5	7	9
1	7	8	3	5	9	6	4	2

Puzzle #28

4	7	2	9	8	3	5	6	1
5	8	9	4	6	1	7	3	2
3	6	1	5	7	2	4	8	9
8	4	6	7	2	9	3	1	5
7	9	3	6	1	5	2	4	8
2	1	5	8	3	4	6	9	7
9	5	7	3	4	8	1	2	6
1	3	8	2	5	6	9	7	4
6	2	4	1	9	7	8	5	3

Puzzle #29

1	4	5	6	8	2	3	9	7
3	9	7	5	4	1	2	8	6
8	6	2	7	9	3	4	1	5
2	8	4	1	6	9	5	7	3
7	3	6	2	5	8	9	4	1
5	1	9	3	7	4	6	2	8
9	7	1	4	3	5	8	6	2
4	2	3	8	1	6	7	5	9
6	5	8	9	2	7	1	3	4

Puzzle #30

4	9	8	2	5	3	1	6	7
6	2	7	9	4	1	3	8	5
1	5	3	8	7	6	2	9	4
7	8	5	6	3	9	4	1	2
9	6	2	7	1	4	5	3	8
3	4	1	5	2	8	9	7	6
5	7	9	3	8	2	6	4	1
8	1	6	4	9	5	7	2	3
2	3	4	1	6	7	8	5	9

Puzzle #31

5	6	3	8	1	7	2	9	4
9	7	2	3	5	4	1	8	6
8	1	4	2	9	6	3	5	7
3	2	6	1	4	8	9	7	5
4	5	8	7	3	9	6	1	2
7	9	1	6	2	5	8	4	3
2	8	7	5	6	1	4	3	9
1	3	9	4	7	2	5	6	8
6	4	5	9	8	3	7	2	1

Puzzle #32

8	7	4	3	6	9	1	2	5
9	2	3	1	5	4	8	7	6
5	6	1	8	7	2	3	4	9
3	1	7	4	2	5	6	9	8
4	9	2	6	1	8	5	3	7
6	8	5	9	3	7	2	1	4
7	4	6	2	8	1	9	5	3
2	5	8	7	9	3	4	6	1
1	3	9	5	4	6	7	8	2

Puzzle #33

3	7	4	8	5	2	1	9	6
1	6	5	4	3	9	7	2	8
2	8	9	1	6	7	5	3	4
9	2	7	5	8	4	6	1	3
8	3	1	7	9	6	2	4	5
5	4	6	3	2	1	8	7	9
4	9	8	2	1	5	3	6	7
6	5	2	9	7	3	4	8	1
7	1	3	6	4	8	9	5	2

Puzzle #34

1	9	8	3	4	6	7	2	5
2	7	5	9	1	8	6	4	3
6	4	3	2	5	7	9	1	8
4	8	1	7	3	5	2	9	6
9	5	6	1	8	2	3	7	4
7	3	2	6	9	4	5	8	1
5	2	9	8	6	1	4	3	7
8	6	7	4	2	3	1	5	9
3	1	4	5	7	9	8	6	2

Puzzle #35

1	9	5	2	8	6	7	3	4
4	6	8	9	7	3	5	2	1
7	2	3	5	4	1	9	8	6
5	3	6	7	9	4	8	1	2
2	7	1	6	5	8	3	4	9
9	8	4	1	3	2	6	5	7
8	5	2	4	6	9	1	7	3
6	1	7	3	2	5	4	9	8
3	4	9	8	1	7	2	6	5

Puzzle #36

4	2	1	9	8	7	5	3	6
8	3	6	1	4	5	7	2	9
7	5	9	2	6	3	1	8	4
5	7	4	3	1	6	8	9	2
9	1	2	8	7	4	3	6	5
3	6	8	5	2	9	4	1	7
1	8	7	6	5	2	9	4	3
2	4	3	7	9	1	6	5	8
6	9	5	4	3	8	2	7	1

Puzzle #37

2	6	4	5	9	1	3	7	8
3	8	7	2	4	6	5	1	9
9	5	1	7	3	8	6	4	2
6	4	9	8	1	7	2	5	3
1	2	5	3	6	9	4	8	7
7	3	8	4	5	2	1	9	6
5	9	6	1	7	3	8	2	4
8	1	3	9	2	4	7	6	5
4	7	2	6	8	5	9	3	1

Puzzle #38

7	2	5	1	8	3	9	6	4
4	6	9	5	7	2	1	8	3
8	3	1	9	6	4	7	2	5
9	4	2	3	5	6	8	7	1
3	7	6	2	1	8	5	4	9
5	1	8	7	4	9	6	3	2
6	8	3	4	9	5	2	1	7
2	5	7	8	3	1	4	9	6
1	9	4	6	2	7	3	5	8

Puzzle #39

3	4	5	1	8	7	2	9	6
6	9	1	3	4	2	8	7	5
7	2	8	6	5	9	1	3	4
2	6	3	4	7	1	9	5	8
1	8	9	5	2	3	4	6	7
4	5	7	9	6	8	3	2	1
8	7	6	2	9	4	5	1	3
9	1	4	7	3	5	6	8	2
5	3	2	8	1	6	7	4	9

Puzzle #40

6	2	1	8	5	9	4	3	7
9	3	7	1	6	4	5	2	8
8	4	5	7	2	3	1	6	9
4	8	6	5	3	7	9	1	2
7	9	3	2	4	1	6	8	5
1	5	2	9	8	6	3	7	4
3	7	8	6	9	5	2	4	1
5	1	4	3	7	2	8	9	6
2	6	9	4	1	8	7	5	3

Puzzle #41

8	5	7	4	2	1	3	9	6
1	2	6	9	3	8	5	7	4
4	9	3	6	7	5	1	2	8
3	7	5	2	6	9	4	8	1
2	6	4	1	8	7	9	5	3
9	8	1	3	5	4	2	6	7
7	3	8	5	4	2	6	1	9
6	1	2	7	9	3	8	4	5
5	4	9	8	1	6	7	3	2

Puzzle #42

7	8	4	5	6	2	1	9	3
1	5	6	9	7	3	2	4	8
3	2	9	8	1	4	5	7	6
8	3	1	2	5	9	7	6	4
9	4	5	7	8	6	3	1	2
6	7	2	3	4	1	9	8	5
4	1	3	6	9	5	8	2	7
2	9	7	4	3	8	6	5	1
5	6	8	1	2	7	4	3	9

Puzzle #43

4	5	6	8	3	2	9	7	1
3	7	1	9	4	5	6	2	8
9	8	2	6	7	1	3	4	5
5	2	9	4	8	6	1	3	7
8	1	3	7	5	9	4	6	2
7	6	4	1	2	3	5	8	9
1	4	5	2	6	8	7	9	3
6	9	8	3	1	7	2	5	4
2	3	7	5	9	4	8	1	6

Puzzle #44

8	7	6	4	1	5	3	9	2
2	4	1	7	9	3	6	5	8
3	5	9	6	8	2	7	1	4
5	2	8	3	4	1	9	7	6
7	1	3	9	2	6	4	8	5
9	6	4	8	5	7	1	2	3
4	9	2	1	3	8	5	6	7
6	3	5	2	7	9	8	4	1
1	8	7	5	6	4	2	3	9

Puzzle #45

9	2	5	7	8	1	4	6	3
4	1	6	9	3	2	7	5	8
7	3	8	5	4	6	1	2	9
2	8	4	3	7	5	6	9	1
1	6	7	2	9	4	8	3	5
5	9	3	1	6	8	2	7	4
3	7	1	8	2	9	5	4	6
8	4	2	6	5	3	9	1	7
6	5	9	4	1	7	3	8	2

Puzzle #46

7	8	4	5	3	2	1	6	9
3	1	2	9	8	6	5	4	7
5	6	9	1	4	7	8	2	3
4	5	1	7	2	3	9	8	6
9	7	3	6	5	8	2	1	4
6	2	8	4	1	9	3	7	5
2	4	7	3	9	1	6	5	8
1	9	5	8	6	4	7	3	2
8	3	6	2	7	5	4	9	1

Puzzle #47

3	5	6	4	7	8	1	2	9
9	1	4	6	2	5	7	3	8
8	7	2	9	3	1	4	6	5
5	4	1	2	8	6	3	9	7
7	9	3	5	1	4	6	8	2
6	2	8	3	9	7	5	4	1
2	3	5	1	4	9	8	7	6
4	6	7	8	5	2	9	1	3
1	8	9	7	6	3	2	5	4

Puzzle #48

2	9	4	6	3	8	1	7	5
5	3	1	9	4	7	6	8	2
6	7	8	1	2	5	4	9	3
3	6	9	4	7	1	2	5	8
1	5	7	8	9	2	3	6	4
8	4	2	5	6	3	7	1	9
4	2	5	7	1	9	8	3	6
7	8	3	2	5	6	9	4	1
9	1	6	3	8	4	5	2	7

Puzzle #49

2	4	9	3	1	8	7	6	5
1	7	8	2	5	6	9	3	4
3	5	6	7	4	9	1	8	2
9	6	1	8	2	5	3	4	7
8	3	4	9	6	7	5	2	1
7	2	5	1	3	4	6	9	8
4	8	3	5	9	1	2	7	6
6	1	2	4	7	3	8	5	9
5	9	7	6	8	2	4	1	3

Puzzle #50

1	5	2	6	8	7	3	4	9
7	4	3	2	9	1	5	8	6
6	9	8	4	3	5	7	1	2
2	1	5	7	6	8	4	9	3
9	6	7	5	4	3	1	2	8
3	8	4	1	2	9	6	7	5
8	2	6	3	7	4	9	5	1
5	7	9	8	1	6	2	3	4
4	3	1	9	5	2	8	6	7

Puzzle #51

8	5	4	1	6	9	3	2	7
1	3	6	4	2	7	9	8	5
9	2	7	8	3	5	4	1	6
2	7	3	6	9	8	5	4	1
6	9	1	3	5	4	2	7	8
4	8	5	7	1	2	6	9	3
5	4	8	2	7	3	1	6	9
3	6	2	9	8	1	7	5	4
7	1	9	5	4	6	8	3	2

Puzzle #52

8	2	9	5	1	3	4	6	7
7	5	4	2	8	6	3	9	1
6	3	1	4	9	7	8	5	2
9	8	5	7	3	4	1	2	6
3	4	7	1	6	2	5	8	9
1	6	2	8	5	9	7	4	3
4	7	8	6	2	1	9	3	5
5	9	6	3	7	8	2	1	4
2	1	3	9	4	5	6	7	8

Puzzle #53

7	6	5	3	8	9	2	1	4
8	9	1	7	2	4	6	3	5
2	3	4	6	5	1	7	8	9
5	1	6	4	7	3	9	2	8
3	8	9	5	6	2	4	7	1
4	7	2	9	1	8	5	6	3
9	4	8	2	3	7	1	5	6
1	5	7	8	9	6	3	4	2
6	2	3	1	4	5	8	9	7

Puzzle #54

1	5	2	8	4	7	3	9	6
6	4	3	5	2	9	1	8	7
9	7	8	3	1	6	2	4	5
7	3	5	6	9	1	8	2	4
4	2	6	7	8	3	9	5	1
8	1	9	4	5	2	6	7	3
5	8	1	2	6	4	7	3	9
2	6	7	9	3	5	4	1	8
3	9	4	1	7	8	5	6	2

Puzzle #55

1	2	4	5	9	8	3	7	6
8	6	5	4	3	7	9	1	2
7	9	3	1	6	2	8	5	4
5	8	9	2	4	3	1	6	7
4	7	6	8	1	9	5	2	3
3	1	2	7	5	6	4	9	8
6	4	8	9	2	1	7	3	5
2	5	1	3	7	4	6	8	9
9	3	7	6	8	5	2	4	1

Puzzle #56

7	1	2	3	4	5	9	8	6
9	6	4	8	1	2	3	7	5
3	8	5	6	7	9	1	4	2
5	2	7	9	3	1	4	6	8
6	9	3	4	5	8	7	2	1
8	4	1	7	2	6	5	3	9
2	3	8	1	9	7	6	5	4
4	5	9	2	6	3	8	1	7
1	7	6	5	8	4	2	9	3

Puzzle #57

2	8	5	3	1	7	4	6	9
7	6	1	5	4	9	2	3	8
3	4	9	6	8	2	1	7	5
5	1	6	2	9	4	7	8	3
4	3	8	1	7	6	5	9	2
9	7	2	8	3	5	6	4	1
8	2	4	9	6	1	3	5	7
6	5	3	7	2	8	9	1	4
1	9	7	4	5	3	8	2	6

Puzzle #58

2	5	7	9	1	6	3	8	4
9	6	8	2	4	3	7	5	1
4	3	1	5	8	7	6	2	9
5	1	4	3	9	2	8	7	6
7	8	6	1	5	4	2	9	3
3	2	9	7	6	8	1	4	5
6	9	5	8	7	1	4	3	2
1	7	3	4	2	5	9	6	8
8	4	2	6	3	9	5	1	7

Puzzle #59

2	9	6	1	7	3	8	5	4
7	4	5	2	9	8	6	3	1
3	8	1	6	5	4	9	7	2
1	5	9	4	2	6	7	8	3
4	3	7	9	8	5	1	2	6
6	2	8	7	3	1	5	4	9
9	1	3	8	4	7	2	6	5
8	6	4	5	1	2	3	9	7
5	7	2	3	6	9	4	1	8

Puzzle #60

7	8	2	5	4	1	3	6	9
4	9	5	2	3	6	7	1	8
6	1	3	7	8	9	4	5	2
3	7	6	1	5	2	8	9	4
1	5	9	4	7	8	6	2	3
8	2	4	6	9	3	5	7	1
9	3	7	8	1	5	2	4	6
2	4	1	3	6	7	9	8	5
5	6	8	9	2	4	1	3	7

Puzzle #61

4	6	3	5	2	1	9	7	8
2	1	5	7	8	9	6	4	3
9	8	7	6	4	3	2	5	1
1	4	6	9	7	8	5	3	2
3	5	9	2	1	6	4	8	7
7	2	8	4	3	5	1	9	6
5	9	2	3	6	7	8	1	4
8	3	4	1	5	2	7	6	9
6	7	1	8	9	4	3	2	5

Puzzle #62

8	1	3	6	7	5	2	4	9
5	7	9	3	2	4	6	1	8
4	2	6	9	8	1	5	3	7
9	4	7	8	6	3	1	5	2
6	8	5	7	1	2	4	9	3
1	3	2	4	5	9	7	8	6
7	9	4	1	3	6	8	2	5
2	6	1	5	9	8	3	7	4
3	5	8	2	4	7	9	6	1

Puzzle #63

9	3	8	4	1	6	7	5	2
6	7	4	2	3	5	1	8	9
2	5	1	7	9	8	4	6	3
1	6	7	9	4	2	5	3	8
3	2	9	8	5	1	6	4	7
8	4	5	6	7	3	9	2	1
7	9	6	3	8	4	2	1	5
4	1	3	5	2	9	8	7	6
5	8	2	1	6	7	3	9	4

Puzzle #64

7	5	2	6	9	8	3	4	1
4	9	8	2	1	3	5	7	6
6	3	1	4	5	7	8	2	9
9	8	5	1	2	6	7	3	4
2	6	4	3	7	5	9	1	8
3	1	7	9	8	4	2	6	5
8	4	9	7	6	2	1	5	3
5	7	3	8	4	1	6	9	2
1	2	6	5	3	9	4	8	7

Puzzle #65

9	7	1	2	6	8	5	3	4
8	6	2	4	3	5	1	7	9
5	4	3	1	9	7	2	8	6
3	1	5	8	7	9	6	4	2
2	9	6	3	5	4	8	1	7
7	8	4	6	2	1	9	5	3
4	5	7	9	1	6	3	2	8
1	2	9	7	8	3	4	6	5
6	3	8	5	4	2	7	9	1

Puzzle #66

9	8	5	2	4	3	7	6	1
1	7	2	9	5	6	8	4	3
6	3	4	7	8	1	9	2	5
5	4	6	1	7	2	3	8	9
8	2	1	5	3	9	6	7	4
3	9	7	8	6	4	5	1	2
4	6	9	3	1	8	2	5	7
2	5	8	4	9	7	1	3	6
7	1	3	6	2	5	4	9	8

Puzzle #67

5	8	1	9	4	7	2	3	6
7	9	3	8	6	2	1	4	5
2	6	4	3	1	5	7	9	8
6	1	9	4	3	8	5	7	2
8	3	2	7	5	9	6	1	4
4	7	5	6	2	1	3	8	9
1	4	8	2	7	6	9	5	3
3	5	6	1	9	4	8	2	7
9	2	7	5	8	3	4	6	1

Puzzle #68

7	4	6	2	8	5	1	9	3
2	3	5	1	6	9	7	8	4
9	1	8	4	3	7	5	2	6
5	6	4	3	7	2	9	1	8
3	9	2	8	5	1	6	4	7
1	8	7	6	9	4	3	5	2
6	5	9	7	4	8	2	3	1
8	7	1	5	2	3	4	6	9
4	2	3	9	1	6	8	7	5

Puzzle #69

3	1	9	5	6	2	8	4	7
5	2	4	9	7	8	6	1	3
7	8	6	1	4	3	5	9	2
1	3	8	6	2	4	9	7	5
9	7	2	3	1	5	4	8	6
6	4	5	8	9	7	2	3	1
4	5	7	2	3	9	1	6	8
8	6	3	4	5	1	7	2	9
2	9	1	7	8	6	3	5	4

Puzzle #70

4	3	2	1	7	9	5	8	6
8	6	5	4	3	2	9	7	1
9	1	7	8	5	6	2	4	3
7	2	4	3	6	8	1	9	5
1	8	6	7	9	5	4	3	2
5	9	3	2	1	4	7	6	8
3	7	8	9	2	1	6	5	4
2	5	9	6	4	3	8	1	7
6	4	1	5	8	7	3	2	9

Puzzle #71

7	6	5	2	1	9	3	4	8
3	2	1	8	7	4	9	5	6
4	8	9	6	5	3	7	1	2
8	9	7	4	6	5	2	3	1
5	3	6	1	9	2	4	8	7
1	4	2	3	8	7	6	9	5
9	5	3	7	2	8	1	6	4
2	1	8	9	4	6	5	7	3
6	7	4	5	3	1	8	2	9

Puzzle #72

9	7	1	6	4	8	5	2	3
4	2	5	3	9	1	7	6	8
3	8	6	5	2	7	1	4	9
7	1	2	4	6	9	8	3	5
5	9	3	1	8	2	6	7	4
8	6	4	7	5	3	2	9	1
6	3	9	2	1	5	4	8	7
1	4	7	8	3	6	9	5	2
2	5	8	9	7	4	3	1	6

Puzzle #73

8	9	3	6	1	5	4	2	7
6	4	1	7	8	2	3	9	5
7	5	2	3	9	4	1	8	6
4	1	5	2	6	9	8	7	3
9	7	8	4	3	1	5	6	2
2	3	6	8	5	7	9	4	1
5	8	7	1	4	6	2	3	9
1	6	4	9	2	3	7	5	8
3	2	9	5	7	8	6	1	4

Puzzle #74

8	9	7	6	5	1	2	3	4
6	2	4	8	9	3	5	7	1
1	3	5	7	4	2	9	6	8
2	5	8	9	7	4	3	1	6
9	7	6	1	3	8	4	2	5
4	1	3	5	2	6	7	8	9
7	4	1	3	8	5	6	9	2
5	8	9	2	6	7	1	4	3
3	6	2	4	1	9	8	5	7

Puzzle #75

5	2	1	6	4	8	9	3	7
8	3	7	9	2	5	6	1	4
4	9	6	1	3	7	2	8	5
3	4	9	2	7	6	8	5	1
1	8	2	3	5	9	7	4	6
6	7	5	4	8	1	3	2	9
7	1	3	5	9	2	4	6	8
2	5	8	7	6	4	1	9	3
9	6	4	8	1	3	5	7	2

Puzzle #76

2	4	5	7	3	8	6	9	1
8	1	6	2	4	9	7	3	5
3	9	7	1	6	5	4	2	8
5	6	2	4	7	1	9	8	3
4	7	8	9	2	3	5	1	6
9	3	1	8	5	6	2	7	4
7	2	3	5	8	4	1	6	9
6	5	9	3	1	2	8	4	7
1	8	4	6	9	7	3	5	2

Puzzle #77

3	8	7	6	5	9	2	4	1
2	9	6	7	4	1	3	5	8
4	1	5	2	3	8	6	9	7
7	5	3	4	9	2	8	1	6
1	2	9	3	8	6	4	7	5
6	4	8	5	1	7	9	3	2
9	6	1	8	7	4	5	2	3
5	7	2	9	6	3	1	8	4
8	3	4	1	2	5	7	6	9

Puzzle #78

6	2	4	9	3	1	8	5	7
8	3	1	4	5	7	2	9	6
9	5	7	2	8	6	4	3	1
7	8	6	1	9	5	3	2	4
5	4	2	6	7	3	1	8	9
1	9	3	8	4	2	7	6	5
4	1	9	5	2	8	6	7	3
3	6	8	7	1	9	5	4	2
2	7	5	3	6	4	9	1	8

Puzzle #79

9	5	3	2	6	8	1	7	4
2	6	7	3	4	1	5	9	8
4	8	1	5	7	9	3	2	6
7	1	5	4	8	3	9	6	2
6	4	2	7	9	5	8	1	3
3	9	8	6	1	2	7	4	5
5	2	6	9	3	7	4	8	1
1	3	9	8	2	4	6	5	7
8	7	4	1	5	6	2	3	9

Puzzle #80

3	1	2	9	5	8	7	6	4
6	4	8	2	3	7	1	5	9
9	7	5	4	6	1	2	8	3
8	5	3	7	4	9	6	2	1
1	2	6	3	8	5	4	9	7
4	9	7	1	2	6	5	3	8
5	8	9	6	1	4	3	7	2
7	3	4	5	9	2	8	1	6
2	6	1	8	7	3	9	4	5

Puzzle #81

7	4	6	2	9	5	3	8	1
1	9	8	3	7	6	4	2	5
5	3	2	8	1	4	6	7	9
6	5	1	7	3	9	8	4	2
2	7	3	1	4	8	5	9	6
9	8	4	6	5	2	7	1	3
3	1	5	4	2	7	9	6	8
8	2	7	9	6	3	1	5	4
4	6	9	5	8	1	2	3	7

Puzzle #82

1	4	3	2	6	7	5	9	8
8	9	6	4	5	1	7	3	2
5	2	7	8	9	3	6	1	4
9	7	5	6	3	8	4	2	1
3	8	4	1	7	2	9	5	6
2	6	1	9	4	5	8	7	3
4	3	2	5	8	9	1	6	7
7	5	8	3	1	6	2	4	9
6	1	9	7	2	4	3	8	5

Puzzle #83

1	2	9	4	8	6	7	3	5
7	6	5	1	9	3	8	4	2
4	8	3	2	7	5	6	1	9
8	1	4	7	6	9	2	5	3
9	3	7	8	5	2	4	6	1
6	5	2	3	1	4	9	8	7
2	7	1	6	3	8	5	9	4
5	4	6	9	2	1	3	7	8
3	9	8	5	4	7	1	2	6

Puzzle #84

4	5	8	6	7	2	9	3	1
3	1	9	8	5	4	6	2	7
7	6	2	3	9	1	8	4	5
1	7	3	2	4	8	5	9	6
9	2	6	5	1	3	4	7	8
5	8	4	7	6	9	3	1	2
2	3	1	4	8	6	7	5	9
6	9	7	1	3	5	2	8	4
8	4	5	9	2	7	1	6	3

Puzzle #85

3	5	2	8	9	7	1	6	4
1	8	7	6	3	4	5	2	9
4	6	9	2	5	1	7	8	3
5	7	1	9	6	8	3	4	2
6	3	8	7	4	2	9	1	5
2	9	4	5	1	3	6	7	8
9	1	3	4	2	6	8	5	7
8	4	6	3	7	5	2	9	1
7	2	5	1	8	9	4	3	6

Puzzle #86

3	7	8	2	9	4	5	6	1
2	5	9	1	8	6	4	7	3
4	1	6	5	7	3	2	9	8
9	2	7	6	1	5	8	3	4
1	3	4	7	2	8	6	5	9
8	6	5	3	4	9	7	1	2
7	8	2	9	5	1	3	4	6
6	4	1	8	3	7	9	2	5
5	9	3	4	6	2	1	8	7

Puzzle #87

4	3	6	7	5	8	1	2	9
8	2	7	9	1	6	3	4	5
5	1	9	4	3	2	8	6	7
9	8	2	6	4	5	7	1	3
3	5	1	8	2	7	6	9	4
6	7	4	3	9	1	5	8	2
2	6	5	1	7	9	4	3	8
1	9	3	5	8	4	2	7	6
7	4	8	2	6	3	9	5	1

Puzzle #88

3	2	1	7	9	4	5	8	6
8	5	6	2	1	3	7	4	9
4	7	9	6	5	8	3	1	2
1	9	4	5	8	7	2	6	3
5	8	3	1	2	6	9	7	4
7	6	2	4	3	9	8	5	1
6	3	7	8	4	2	1	9	5
2	4	5	9	7	1	6	3	8
9	1	8	3	6	5	4	2	7

Puzzle #89

2	3	4	1	5	8	7	9	6
5	8	9	7	2	6	4	3	1
1	7	6	3	4	9	8	2	5
9	4	5	8	7	2	6	1	3
3	6	7	4	1	5	2	8	9
8	1	2	9	6	3	5	7	4
7	5	8	6	9	1	3	4	2
6	9	3	2	8	4	1	5	7
4	2	1	5	3	7	9	6	8

Puzzle #90

3	1	5	7	4	6	8	9	2
8	2	7	3	5	9	1	6	4
6	9	4	8	2	1	5	3	7
1	6	8	2	9	4	7	5	3
9	5	3	1	7	8	2	4	6
4	7	2	5	6	3	9	1	8
5	4	1	6	8	2	3	7	9
7	8	9	4	3	5	6	2	1
2	3	6	9	1	7	4	8	5

Puzzle #91

7	4	9	1	2	8	6	3	5
5	2	8	3	9	6	1	7	4
3	1	6	4	7	5	2	8	9
8	6	5	2	1	7	9	4	3
9	7	2	6	3	4	8	5	1
4	3	1	5	8	9	7	6	2
2	8	3	7	5	1	4	9	6
1	9	4	8	6	3	5	2	7
6	5	7	9	4	2	3	1	8

Puzzle #92

9	7	4	3	2	8	5	1	6
2	5	8	1	6	7	4	3	9
3	6	1	9	4	5	7	8	2
7	9	2	6	5	3	8	4	1
1	3	5	2	8	4	6	9	7
4	8	6	7	9	1	3	2	5
8	1	9	4	7	6	2	5	3
5	2	7	8	3	9	1	6	4
6	4	3	5	1	2	9	7	8

Puzzle #93

5	4	8	7	6	9	2	3	1
1	2	7	3	8	5	9	4	6
6	9	3	4	1	2	5	8	7
3	6	1	5	4	7	8	9	2
9	5	4	8	2	1	6	7	3
7	8	2	6	9	3	4	1	5
4	1	5	2	3	8	7	6	9
2	3	6	9	7	4	1	5	8
8	7	9	1	5	6	3	2	4

Puzzle #94

5	3	7	2	9	4	6	1	8
9	4	6	7	8	1	5	3	2
2	8	1	6	3	5	7	4	9
1	2	9	5	4	3	8	7	6
4	7	3	9	6	8	1	2	5
6	5	8	1	7	2	3	9	4
3	1	2	8	5	9	4	6	7
8	6	4	3	2	7	9	5	1
7	9	5	4	1	6	2	8	3

Puzzle #95

6	1	8	4	3	5	9	7	2
4	5	7	2	9	6	8	1	3
3	2	9	8	7	1	6	4	5
7	4	2	3	5	8	1	9	6
9	6	1	7	2	4	3	5	8
8	3	5	6	1	9	4	2	7
1	8	6	5	4	7	2	3	9
2	7	4	9	8	3	5	6	1
5	9	3	1	6	2	7	8	4

Puzzle #96

2	3	8	7	6	5	1	4	9
9	1	6	4	2	3	8	7	5
4	7	5	9	8	1	3	6	2
8	4	9	6	1	2	7	5	3
5	6	7	8	3	4	9	2	1
3	2	1	5	9	7	6	8	4
1	5	4	3	7	8	2	9	6
6	8	2	1	5	9	4	3	7
7	9	3	2	4	6	5	1	8

Puzzle #97

1	5	6	8	4	2	7	3	9
8	3	7	5	6	9	4	1	2
9	2	4	3	1	7	5	6	8
7	6	8	4	9	5	3	2	1
5	4	2	1	3	8	9	7	6
3	9	1	2	7	6	8	5	4
2	8	9	6	5	3	1	4	7
6	1	3	7	8	4	2	9	5
4	7	5	9	2	1	6	8	3

Puzzle #98

7	6	4	1	8	3	9	5	2
8	9	1	4	2	5	3	6	7
5	2	3	6	7	9	8	4	1
1	3	8	5	6	2	4	7	9
2	7	9	8	4	1	6	3	5
6	4	5	3	9	7	1	2	8
9	8	7	2	3	6	5	1	4
4	1	6	7	5	8	2	9	3
3	5	2	9	1	4	7	8	6

Puzzle #99

7	4	8	1	5	9	2	6	3
3	5	2	4	7	6	8	9	1
9	1	6	2	3	8	5	7	4
1	8	9	6	2	7	3	4	5
6	2	3	9	4	5	7	1	8
4	7	5	8	1	3	6	2	9
5	3	1	7	6	4	9	8	2
2	9	7	5	8	1	4	3	6
8	6	4	3	9	2	1	5	7

Puzzle #100

6	4	5	2	3	1	7	8	9
8	2	3	4	9	7	1	6	5
9	1	7	6	8	5	2	3	4
5	8	6	3	4	2	9	7	1
2	3	9	7	1	6	5	4	8
4	7	1	8	5	9	6	2	3
7	5	8	9	2	4	3	1	6
3	9	2	1	6	8	4	5	7
1	6	4	5	7	3	8	9	2

Puzzle #101

9	3	8	4	7	1	6	2	5
1	2	4	9	5	6	7	8	3
5	6	7	8	2	3	1	9	4
8	1	5	7	6	9	4	3	2
3	4	6	2	1	5	9	7	8
7	9	2	3	8	4	5	6	1
6	5	9	1	3	2	8	4	7
2	8	1	6	4	7	3	5	9
4	7	3	5	9	8	2	1	6

Puzzle #102

3	8	9	2	4	5	7	6	1
4	5	7	8	6	1	2	9	3
6	2	1	7	9	3	4	8	5
8	3	2	6	1	4	9	5	7
7	9	6	3	5	2	8	1	4
5	1	4	9	8	7	6	3	2
1	4	8	5	2	6	3	7	9
9	7	5	4	3	8	1	2	6
2	6	3	1	7	9	5	4	8

Puzzle #103

9	2	8	1	6	4	7	3	5
6	5	1	9	7	3	2	8	4
4	7	3	2	5	8	1	6	9
5	3	4	6	2	7	8	9	1
8	6	9	3	4	1	5	7	2
7	1	2	8	9	5	6	4	3
3	9	6	5	8	2	4	1	7
2	8	7	4	1	9	3	5	6
1	4	5	7	3	6	9	2	8

Puzzle #104

3	8	9	5	7	6	4	2	1
6	7	1	3	2	4	5	9	8
2	5	4	8	1	9	3	6	7
4	9	3	1	6	8	2	7	5
5	6	7	2	4	3	1	8	9
8	1	2	7	9	5	6	3	4
9	4	8	6	3	1	7	5	2
1	2	6	9	5	7	8	4	3
7	3	5	4	8	2	9	1	6

Puzzle #105

4	5	3	6	8	2	1	7	9
2	9	6	3	7	1	4	5	8
7	1	8	9	5	4	3	2	6
9	2	1	8	6	3	5	4	7
8	3	7	4	2	5	6	9	1
5	6	4	7	1	9	8	3	2
3	4	2	1	9	8	7	6	5
1	7	5	2	3	6	9	8	4
6	8	9	5	4	7	2	1	3

Puzzle #106

7	8	6	9	2	5	4	1	3
1	9	5	8	4	3	6	2	7
3	4	2	1	6	7	5	8	9
2	3	4	7	9	8	1	5	6
9	7	1	2	5	6	8	3	4
6	5	8	3	1	4	9	7	2
5	6	3	4	7	1	2	9	8
8	1	9	6	3	2	7	4	5
4	2	7	5	8	9	3	6	1

Puzzle #107

9	7	4	5	2	6	1	3	8
6	1	8	7	4	3	5	9	2
5	2	3	8	9	1	6	7	4
1	9	6	4	3	5	8	2	7
2	3	5	9	7	8	4	6	1
8	4	7	6	1	2	9	5	3
4	6	2	3	8	9	7	1	5
3	8	9	1	5	7	2	4	6
7	5	1	2	6	4	3	8	9

Puzzle #108

5	7	9	4	1	3	2	6	8
4	1	3	8	2	6	9	5	7
8	6	2	9	7	5	4	1	3
3	8	1	6	5	9	7	2	4
2	5	4	1	3	7	8	9	6
7	9	6	2	4	8	5	3	1
1	2	8	3	9	4	6	7	5
6	3	5	7	8	2	1	4	9
9	4	7	5	6	1	3	8	2

Puzzle #109

5	4	6	3	8	1	9	2	7
8	1	3	2	9	7	4	5	6
7	2	9	4	5	6	8	3	1
9	5	7	1	2	4	6	8	3
1	8	4	9	6	3	2	7	5
3	6	2	8	7	5	1	4	9
6	7	1	5	4	8	3	9	2
4	9	5	6	3	2	7	1	8
2	3	8	7	1	9	5	6	4

Puzzle #110

8	9	6	5	7	4	3	2	1
4	1	3	6	2	8	9	7	5
5	2	7	1	3	9	6	8	4
7	6	8	9	4	2	1	5	3
3	5	9	8	1	7	2	4	6
2	4	1	3	5	6	7	9	8
9	8	2	4	6	1	5	3	7
6	3	4	7	9	5	8	1	2
1	7	5	2	8	3	4	6	9

Puzzle #111

8	4	7	9	3	1	5	2	6
1	5	2	6	4	8	3	7	9
3	6	9	5	2	7	1	8	4
5	1	4	7	6	3	2	9	8
7	2	8	4	1	9	6	3	5
6	9	3	8	5	2	7	4	1
9	7	5	2	8	6	4	1	3
4	8	1	3	7	5	9	6	2
2	3	6	1	9	4	8	5	7

Puzzle #112

4	7	9	2	6	3	1	5	8
1	2	8	4	9	5	6	7	3
6	3	5	1	7	8	4	9	2
9	5	3	8	4	1	7	2	6
8	1	7	3	2	6	5	4	9
2	6	4	9	5	7	3	8	1
3	8	2	7	1	4	9	6	5
5	4	1	6	8	9	2	3	7
7	9	6	5	3	2	8	1	4

Puzzle #113

5	1	3	2	8	7	9	4	6
9	4	2	1	6	5	7	3	8
8	7	6	3	4	9	1	2	5
2	9	8	4	5	6	3	7	1
7	6	4	8	1	3	2	5	9
3	5	1	9	7	2	8	6	4
4	3	9	5	2	1	6	8	7
1	8	7	6	3	4	5	9	2
6	2	5	7	9	8	4	1	3

Puzzle #114

1	4	2	9	6	5	3	7	8
3	5	7	4	2	8	9	1	6
9	8	6	7	3	1	2	5	4
7	3	8	5	4	2	6	9	1
5	6	1	3	9	7	4	8	2
4	2	9	8	1	6	7	3	5
2	9	3	1	8	4	5	6	7
8	7	4	6	5	9	1	2	3
6	1	5	2	7	3	8	4	9

Puzzle #115

8	7	6	2	3	5	4	1	9
3	2	5	9	4	1	7	6	8
9	1	4	6	8	7	3	5	2
7	4	9	5	2	8	6	3	1
1	5	8	7	6	3	9	2	4
6	3	2	4	1	9	5	8	7
2	6	3	8	7	4	1	9	5
4	9	1	3	5	2	8	7	6
5	8	7	1	9	6	2	4	3

Puzzle #116

1	7	5	8	6	3	9	4	2
8	2	3	5	4	9	1	7	6
9	4	6	2	1	7	8	3	5
3	8	4	6	7	5	2	1	9
5	9	2	1	3	8	4	6	7
7	6	1	9	2	4	3	5	8
6	5	9	3	8	1	7	2	4
2	3	7	4	9	6	5	8	1
4	1	8	7	5	2	6	9	3

Puzzle #117

5	6	4	2	8	3	7	9	1
9	1	7	6	5	4	2	8	3
8	3	2	9	7	1	5	6	4
3	5	1	8	6	9	4	2	7
2	7	9	3	4	5	8	1	6
4	8	6	7	1	2	3	5	9
6	2	5	4	9	7	1	3	8
1	4	8	5	3	6	9	7	2
7	9	3	1	2	8	6	4	5

Puzzle #118

3	2	6	8	4	7	9	5	1
5	4	1	2	6	9	3	7	8
7	8	9	1	5	3	4	2	6
1	6	8	3	7	4	2	9	5
4	9	3	5	8	2	1	6	7
2	7	5	6	9	1	8	3	4
8	5	4	9	3	6	7	1	2
6	3	2	7	1	8	5	4	9
9	1	7	4	2	5	6	8	3

Puzzle #119

1	5	7	4	2	6	3	8	9
3	4	6	1	8	9	2	5	7
2	8	9	3	7	5	1	4	6
9	7	1	2	6	4	5	3	8
6	3	8	5	1	7	9	2	4
4	2	5	9	3	8	7	6	1
5	9	2	6	4	1	8	7	3
7	1	4	8	5	3	6	9	2
8	6	3	7	9	2	4	1	5

Puzzle #120

4	3	7	9	6	1	2	5	8
5	2	6	8	7	3	1	9	4
1	9	8	5	4	2	7	3	6
7	6	2	4	9	5	8	1	3
8	5	4	1	3	6	9	7	2
9	1	3	7	2	8	6	4	5
3	4	1	2	8	9	5	6	7
6	8	5	3	1	7	4	2	9
2	7	9	6	5	4	3	8	1